THE NE

Gold and Earth's Treasures
EDITION

Oliver Wilde

INSIDE

A word from our editor

HIDDEN under the Earth's surface are treasures. Many people dig, and die, for gold and **precious** minerals. These are natural treasures of the Earth.

They're well hidden and hard to find. That's why they are among the most **valuable** things on Earth.

What will people do for riches? Read on, to find out about the strange and awful things that can happen when people get too greedy.

In this edition of *The News*, you will discover what these treasures are, where they can be found and how they are dug up.

Daily spot

El Dorado (say *ell-da-rar-doe*) is a mythical city that is said to be filled with gold and precious stones. People have searched for it for centuries, but it has never been found.

Treasure map

Where to find the treasure.

EARTH'S MOST valuable treasures are precious metals and precious stones.

For a metal or stone to be precious it must be beautiful, hard to break or wear out, and hard to find.

The most precious metals are gold and silver. Gold has been valuable since **ancient** times. Some countries have no gold left because it was all **mined** hundreds of years ago.

The most precious stones are diamonds, sapphires, rubies and emeralds.

They aren't easy to find. But we at *The News* have made it easy for you to know where to start looking. This map shows where in the world precious metals and stones have been found.

Even though they are hard to find, deposits of precious metals and stones are scattered across nearly every continent in the world.

Fact of the day

South Africa mines about 600 tonnes (590 tons) of gold each year, which is more than any other country in the world.

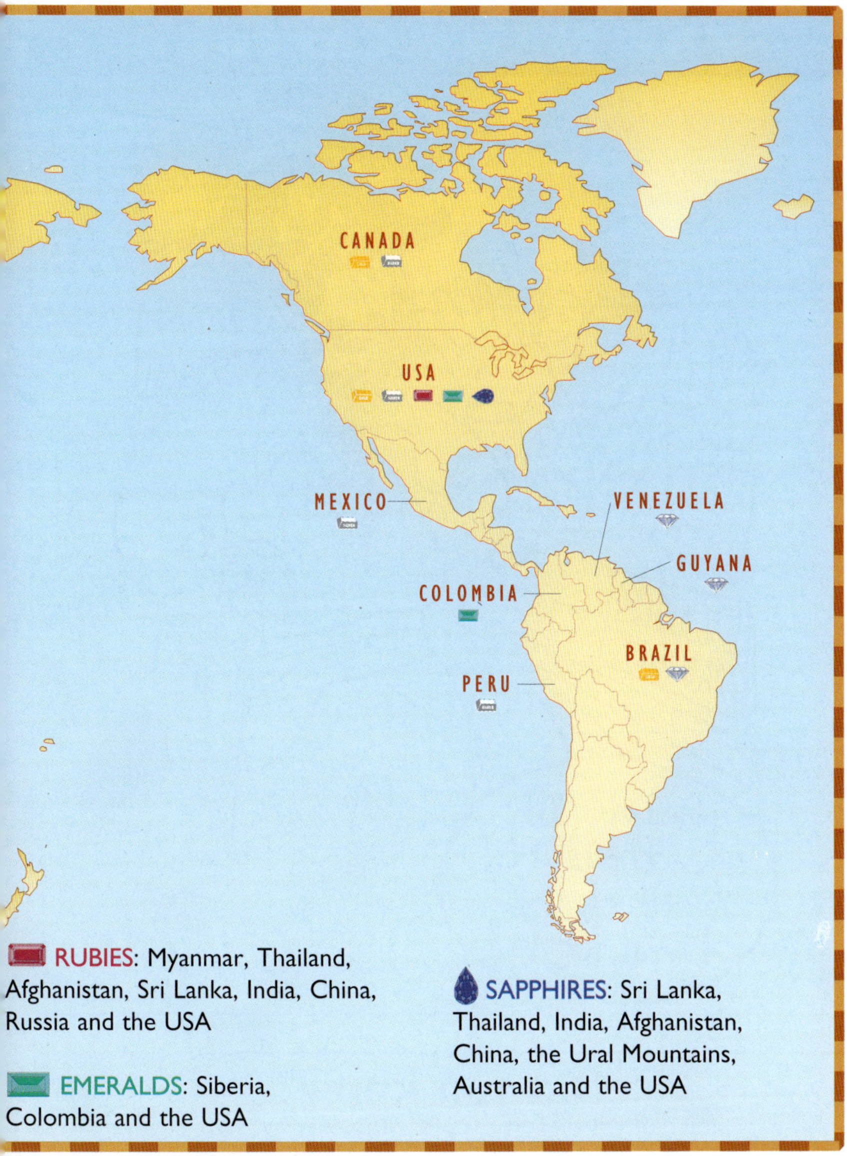

*This map shows the countries where the supplies of precious metals and stones are large enough to be mined for a **profit**.*

Imagine what it would feel like to find buried treasure!

Gold spot

Gold is heavier than any other substance. One cupful of gold weighs 4,630 grams (163 ozs). One cupful of water weighs 240 grams ($8^1/_2$ ozs). The gold is almost 20 times heavier!

Did you know?

Gold can be rolled out so thin that light will shine right through it.

Glittering gold

Earth's most wanted treasure.

SOME PEOPLE love gold more than anything else. They have given up everything to hunt for it. They have left their families. They've even killed and died for gold. It's hard to imagine that anything is that valuable!

To many people, gold is the most valuable of all the Earth's treasures.

People love gold's beautiful colour. It doesn't rust like tin or discolour like silver.

It is soft enough to make into beautiful shapes easily. And it's strong enough to make into fine wire and thin sheets.

Gold isn't easy to find. This helps to make it really valuable.

The News Cartoon

Fact of the day

A piece of gold weighing only 29 grams (1 oz) can be stretched into a wire 100 km (60 miles) long.

Did you know?

The most famous robbers are pirates. But they weren't bank robbers. They roamed the oceans stealing gold and jewels from passenger and trade ships.

Hidden treasure

Mining for gold.

GOLD GATHERS in rocks in the Earth's crust. The crust is the outer layer or skin of the Earth. This crust is between 25 and 90 km (16–56 miles) thick.

Even though gold is found in many countries around the world, most of the **deposits** are very small. If the Earth's outer layer was equal to 10 million tennis balls, only 5 of them would be gold.

Some gold can be found in the surface soil of the Earth in riverbeds and on mountain slopes. It is usually quite easy to get by panning or digging.

But most gold is hidden in rocks deep underground. This gold must be mined.

In underground mines, holes called shafts are dug deep into the Earth. From the shaft, passages are cut into the rock. Then **ore** is blasted out of the rock with explosives and lifted up the shaft to the surface.

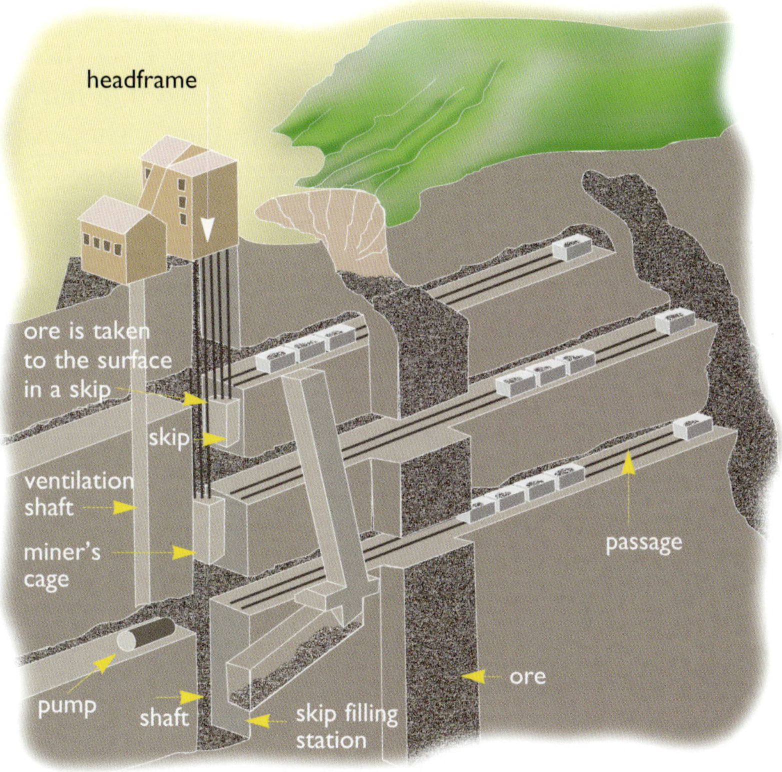

Underground mines must have a ventilation shaft so that the miners receive fresh air. The shaft also stops dangerous gases from building up in the mine.

Word of the day

PANNING—washing soil dug from riverbeds in a round shallow dish to find the gold.

Gold, gold, gold!

Imagine… if you found the diary of a gold digger during the gold rush. You would find out what life was like on a goldfield.

Sunday, 2 July

Arrived in town a few days ago. Most shops were closed with signs on the doors saying "gone to the diggings".

I bought a shovel. The shovel cost twice as much as it did before the gold rush. Then I headed for the diggings myself.

Thursday, 20 July

What madness I see here at every turn. Men in a hurry to find a fortune. They work night and day at their claim. They are stricken by **gold fever**. Some would steal someone else's find rather than go home empty-handed.

Monday, 7 August

I can hardly believe how quickly the camps are growing. Tents crowded round each other and men fighting for space. I had a bit of a find today. I'll not tell anyone lest it get stolen away.

Wednesday, 30 August

I've seen five men strike it rich this very day. Nuggets the size of a fist and worth more money than I've ever seen.

More men arrive each day. Some to search for gold. And some to sell food and equipment to those who can afford to buy it. I am finding more gold every day. I think I'll soon have enough to trade.

Friday, 8 September

Took my gold to be weighed today. Oh, what a madman I am. They beat my small nugget with a hammer and it crumbled into powder.

Another piece they tried to scratch with a knife but it was much too hard. Real gold will flatten, not crumble, they told me. Real gold will scratch easily because it's soft.

It's fool's gold I have and nothing more. Not even my pride. I must return to the diggings and try again to make my fortune.

Fact of the day

The biggest nugget in the world was the Welcome Stranger. It was found in central Victoria on 5 February 1869. John Deason dug the nugget out of the ground with a pick. It weighed 72,028 grams (2,541 ozs).

Words of the day

FOOL'S GOLD—a **mineral** called iron pyrites (say *pie-right-iz*) which looks like gold. Some people are fooled by it and think they have found real gold.

NUGGET—a natural lump of gold.

Diamonds are forever

All that glitters...

DIAMONDS ARE the most precious of all the gemstones. When you think of diamonds, you might think of glittering and sparkling gems. But diamonds are actually made of **carbon** which is the same thing as coal.

It takes an enormous amount of heat and pressure to change carbon into a diamond. The rough stones are formed deep in the Earth. Then they are brought to the surface by the force of erupting volcanoes.

Diamonds are valuable because they are beautiful, **rare**, and very very hard.

Some diamonds are found in the ground. Others are found in rivers, lakes and oceans.

In South Africa, diamonds sometimes wash up on the shores of some beaches. These beaches are often guarded so that the diamonds cannot be stolen.

An expert diamond cutter can turn a rough diamond into a stone of fiery brilliance. Because diamonds are the hardest stone, only diamonds can be used to cut and shape them.

When diamonds are found they look like small dull rocks. These are called rough diamonds. It is only when they are cut and shaped that they look beautiful.

One of the most beautiful things about diamonds is the way they catch and **reflect** light. This makes them look like they are shining.

Hardness scale

Gem experts rate substances according to how hard it is to scratch them. The hardest substance rates a score of 10 and the softest substance rates a score of 1.

diamond	10
sapphire	9
ruby	9
emerald	8
gold	2.5–3
talc	1

(Baby powder is made from talc.)

Did you know?

The largest diamond ever found is called the Cullinan. It was discovered in South Africa in 1905 and weighed 621 grams (22 ozs). It was presented to King Edward VII of Great Britain.

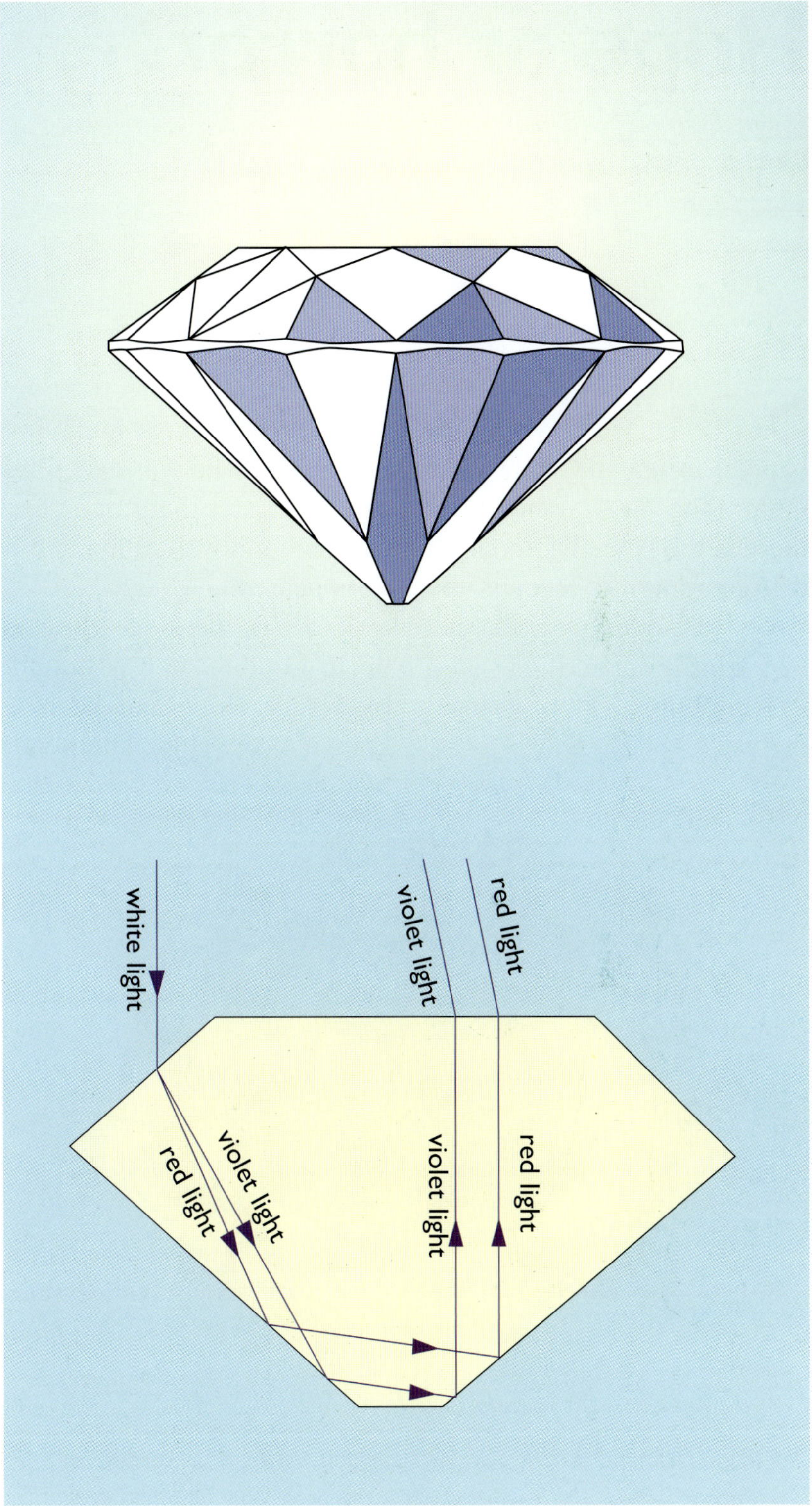

The rough diamond is cut into a shape that has many small flat surfaces. White light bounces off these surfaces and splits into colours that can be seen through the top of the clear stone.

Hope or horror?

Can a gemstone be cursed?

SOME GEMSTONES are said to be cursed. Bad things happen to anyone who owns them. The most famous of these is the Hope Diamond.

According to legend, the stone was stolen from the eye of a **Hindu** statue. It was sold to King Louis XIV of France.

The diamond was worn by many French kings including Louis XVI who was **executed** in 1792. Later that same year, the diamond was stolen and it disappeared.

In 1830, the stone showed up in London. It was bought by Henry Hope and became known as the Hope Diamond.

The Hope Diamond is dark blue in colour and weighs 45.5 carats (9 grams or 1/3 oz). This perfect stone is the finest blue diamond in the world. Blue diamonds get their colour from very small amounts of other minerals in the stone. The Hope Diamond is surrounded by 16 white diamonds and it hangs from a chain that is set with 45 white diamonds.

The diamond had many other owners. Most of them suffered violent deaths and other disasters. This added to the story of the curse.

A European prince bought it and gave it to an actress. He then shot her.

Then a Greek man bought the diamond. His family was killed when their car went over a cliff.

In 1909, the diamond became the property of a Turkish sultan. A few months later, the army led a revolt and took over his country.

In 1911, a rich woman, Evalyn McLean, bought the Hope Diamond. Her husband went insane, her son was killed in a car accident and her daughter killed herself.

When Mrs McLean died, the diamond was sold to a jeweller, Harry Winston. In 1958, Harry gave the Hope Diamond to the Smithsonian Institute in New York. And the curse stopped there.

Sapphires, emeralds, rubies

It's all in the colour.

SAPPHIRES, EMERALDS and rubies are the next most precious stones after diamonds.

Sapphires are produced by erupting volcanoes. That is why they are usually found in or near streams that flow around **extinct** volcanoes. The most prized sapphire is deep blue in colour.

When a sapphire is red, it is called a ruby. There are many different shades of red, but the most valuable one is a deep purplish red known as pigeon blood. The best rubies are found in Myanmar. People have been mining rubies in this part of the world since the Stone Age—that's about 35,000 years altogether!

Emeralds are bright green in colour. Perfect emeralds are worth more than diamonds because they are the hardest gemstone to find.

In ancient times, emeralds were mined in Egypt, where they were thought to be magical. Egyptian people believed that emeralds had the power to heal eye diseases.

This is the mask that the ancient Egyptian ***Pharaoh****, Tutankhamen (say* Toot-arn-car-men*) was buried in when he died. The mask is made of gold and set with many precious gemstones.*

Did you know?

Sapphires are not always blue. They can be yellow, green, gold, violet or brown.

The Black Prince's Ruby

Imagine... if you could talk to a gemstone about its journey from the ground to a crown.

IT ALL BEGAN about six hundred and fifty years ago. I was dug out of the ground from a mine in Ceylon (now known as Sri Lanka). I was placed in the hands of Abu Said, an Arab ruler.

Abu said I was a fabulous ruby. He showed me off to many rich people and I became famous. Then, Abu was invited to dinner at the castle of King Peter the Cruel.

During the main course, King Peter stabbed him to death and then he stole me from Abu's body!

Well, Peter kept me for a while, but I was always well hidden. Until one day the king's brother rebelled. Peter the Cruel called the Black Prince of England for help.

When they won the fight, Peter the Cruel rewarded the English prince for his help. What was the reward? You guessed it—me!

After that, I was known as the Black Prince's Ruby. I was added to the **Crown Jewels** of England and settled into life with the British royalty.

Fact of the day

The Black Prince's Ruby is not actually a ruby. It's a spinel (say spun-ell*), which is a stone that looks like a ruby but isn't as valuable.*

King Henry V wore me in his helmet when he fought the French in 1415. Oh, what a victory that was!

Then I was in Richard III's helmet at the Battle of Bosworth. Er—Richard III died that day. Not my fault.

I was sold to a jeweller in the 1650s, when Oliver Cromwell ruled England. But in 1660, I was back in the Crown Jewels of Charles II.

I've been in a few crowns since then, of course. The first ruling queen to wear me in her crown was Queen Victoria in 1838. I've been in Elizabeth II's crown since 1953. I've come a long way from a rock in the ground!

The Black Prince's Ruby is set in the Imperial Crown of England. It is nestled in the centre of a cross made up of thousands of tiny diamonds. The velvet crown is studded with hundreds of precious gemstones.

Interview with a jeweller

What can you do with gold and gemstones?

Terry, why did you become a jeweller?
Because I get to work with beautiful things every day. I love the sparkle of gems.

What do you use to make jewellery?
I usually use gold and gemstones. The gold comes in three different colours—pink, white and yellow. The gemstones come in all the colours of the rainbow.

Where do you get the gems?
I steal them! No, just kidding. I buy the gems from gem **merchants**. They buy them from mining companies.

What sort of jewellery do you make?
I like to make emerald necklaces because emeralds are my favourite stone. But I also make rings, bangles, earrings and toe rings—whatever the customer wants.

What's the most interesting piece of jewellery you've ever made?
I once made a diamond collar for a cat. Not just an ordinary cat. This cat was owned by a famous movie star. And I made a toe ring for a pop star. It was covered in tiny rubies and looked like a mini crown.

Wow. Who were they?
That's a secret. I can't talk about my customers. They wouldn't come back and I'd be out of a job.

Panning for gold

Follow these steps. You never know what you might find.

PANNING for gold is fun. If you search for gold in or near a creek in an old gold mining area, you really might find a few specks of gold!

If you can't get to an old goldfield, pan for gold in your backyard using a garden hose. You never know what's hiding in the dirt under your feet.

You will need:

- a spade
- a tin plate
- running water

What to do:

1. Take a spade full of dirt and put it into the pan.
2. Hold the pan under gently running water.
3. Make circular motions with the pan so that the dirt swirls around in it.
4. Tip the pan slightly while you swirl it around. Let the water wash the dirt slowly out of the pan.
5. The gold will be left in the pan because it is heavier than dirt.

Myth of gold

The stories of old teach us not to be greedy.

King Midas of Phrygia (say *fridg-ee-ah*) was very powerful. Even the gods granted his wishes.

One day, Midas asked the gods for the power to turn everything that he touched to gold. And he got it.

He ran around touching everything in his throne room. He turned the floor to gold. And the tables. And the chairs. Even the grapes in their bowls turned to gold.

Midas was so excited, he hugged his wife. And yes, she turned to gold too.

Midas was not so excited about that. By dinnertime, Midas was very hungry. But, not even a king can eat food that's turned to gold.

When his bed turned to gold, which was not very comfortable, Midas was quite upset. He begged the gods to take the power back.

The queen was very happy when she turned into a person again. She did not enjoy being a statue.

King Midas learnt a lesson not to be too greedy. After that, he was still very powerful—but wise as well!

Golden gags

A treasure chest of cheer.

True treasure

Glorious, glistening, glittering gold
most precious substance on earth, I'm told
but I'd be a fool to want too much
in case, like Midas, everything I touch
turns to metal.

Splendid, sparkling, shimmering gems
flashing colour in my diadems
but I must be careful to avoid all strife
in case, like Richard, I lose my life
in a battle.

I would like a treasure chest
filled with things that are the best
but none of the treasure in the ground
can make me as happy, I have found,
as a friend will.

a poem by Trisha Isle

JOKES

Q. What kind of jewels does a dead person wear?

A. *Tombstones.*

Q. What do you call gold found in a henhouse?

A. *A chicken nugget.*

Q. Where is the easiest place to find diamonds?

A. *In a pack of cards.*

Q. Which queen never wears a crown?

A. *A queen bee.*

Mining around the world

Gold and gems are not the only treasures on our Earth.

Some of Earth's treasures are more useful than others. Gold and gemstones are beautiful to look at, but they are no more valuable to us than raw materials, such as oil and gas, that we need in our everyday lives. This map shows the countries in which the larger deposits of these useful materials are mined.

Coal is used as fuel. It is burned to give heat or to make the engines in industrial machinery work.

Iron is melted so that it can be mixed with carbon to make steel, which is used to make many things from ships and planes to hairpins and bedsprings.

Gas is stored in huge pipes. It is burned to make heat for homes and industry.

Copper is mined from huge, wide holes that are dug in the ground. It is used for all kinds of electrical wiring and machinery.

Uranium is turned into nuclear energy in power plants like this one. Uranium is used for industry, medicine and weapons.

Offshore oil rigs are built to mine the oil that lies beneath the seabed. We use oil for industry and also for transport.

Editorial

OUR Earth is rich with many different natural treasures. We use rocks and minerals for building, travelling, heating, and making goods such as clothes, furniture and toys.

Experts say that we are using Earth's treasures too quickly. They say that we will run out of materials such as natural gas, oil and gold within 60 years.

Some people think that we should use the sun and water to make the energy we need to live the way we do. That way, the Earth's treasures would last much longer.

At *The News*, we think that everyone should try to preserve Earth's treasures. Use solar power for heating. Walk, cycle or use public transport instead of travelling in a car. And fake jewellery can look as good as the real thing!

Word of the day

ENVIRONMENT—the conditions around us.

Letters

to the editor

Dear Editor,
Mining is bad for the environment. Huge areas of land are cleared of plants. Thousands of tonnes of topsoil are dug up and carted away in trucks. Also, streams can be **polluted** by poisonous chemicals. I think mining should be banned, don't you?

From ENVIRO ERIN

Dear Editor,
I think every country in the world should use gold coins for money. That way we wouldn't have to change our money when we visit other countries. And money would have the same value wherever you go.

From CLEVER CARL

This old copper mine is now just a big hole in the ground. We need to watch out for damage to the environment. We cannot say it's all right to keep on mining until there's nothing left and the land is bare.

Tiara, the treasure expert,

answers your questions

Dear Tiara,
I was reading about gold and found out that it is mined underground. Is it dangerous for the miners?

From CONCERNED CONNIE

TIARA: Underground mining used to be very dangerous. The temperature underground could reach 40°C (104°F). The miners could get lung diseases from the dust, and the mines often caved in.

These days, with proper ventilation and protective clothing and equipment, mines are safer places to work. But, earth movement can still cause cave-ins. Miners can sometimes be hurt or killed if part of a mine caves in while they are working in it.

There will always be a degree of danger for workers in underground mines.

Dear Tiara,
I had my name put on the school captain board at school. The letters are gold-coloured. Are they real gold?

From HOPEFUL HARRY

TIARA: Yes, the letters are real gold. The gold has been beaten to a very thin sheet. It is called gold leaf. The gold-beater uses gold-beater's skin between the leaves of gold while they beat it. The skin is made from a layer of an ox's intestine. Of course, it's been specially treated.

Dear Tiara,
My nanna has gold fillings in her teeth. I think they look really cool. How could I get a gold filling?

From GOLDEN GAIL

TIARA: I've never heard of anyone who actually wanted a filling. I think it's better to look after your teeth properly so you don't need fillings.

But if you were to get a gold filling, the dentist would make a model of your tooth. This is sent to a laboratory where a gold copy of the model is made. The dentist then fits the gold copy onto the tooth.

You might think gold fillings look cool, but they come at a price. They cost six times as much as ordinary fillings.

This building has gold domes. Many years ago, gold was often used to decorate special buildings. Today, gold has increased in value so much that it is too expensive to use as a building material.

Further resources

BOOKS

Your school or local libraries are always a good place to start when looking for information. Look under 620 on the shelves. Here are some titles to start with:

Anna Ciddor *The Goldfields: Through Children's Eyes* Macmillan

John Nicholson *Gold!* Allen & Unwin

R F Symes *Eyewitness: Rocks & Minerals* Dorling Kindersley

NEWSPAPERS

Information about mining and gold and gem values can be found in the financial section of the daily newspaper.

MAGAZINES

There are lots of magazines about gemstones and gem collecting. Check out your local newsagent and see what you can find.

INTERNET

There are plenty of sites about gold and gemstones. Surfing the Net with Kids is a great place to start. For summaries and links to good gold and gemstone sites for kids, go to:
www.surfnetkids.com/goldrush.htm
www.surfnetkids.com/rocks.htm

Another great search engine to try is CyberSleuth Kids at:
www.cybersleuth-kids.com.

Type the words "gold" or "gemstones" into a search engine and you'll find loads of information.

Sovereign Hill, a reconstructed gold mining town at Ballarat in Victoria, has a fantastic web site:
www.sovereignhill.com.au

MUSEUMS

Museums and art galleries often have exhibitions of gemstones and items made from gold. Sometimes they have come from other countries around the world.

Birthstone chart

January	garnet
February	amethyst
March	aquamarine
April	diamond
May	emerald
June	pearl
July	ruby
August	peridot
September	sapphire
October	opal
November	topaz
December	turquoise

When a grain of sand gets into an oyster, a pearl—or three—will grow on the shell.

Glossary

ANCIENT—happened or lived a very long time ago.
CARBON—an element that is found in all living things. Diamonds and coal are made of carbon.
CROWN JEWELS—the jewels that are handed down from one ruling monarch to the next.
DEPOSITS—a layer which collects in a particular place.
EXECUTED—put to death according to a law.
EXTINCT—no longer in existence.
GOLD FEVER—when people were very excited by the thought of finding gold and getting rich it was called "gold fever", because it was like a disease that could not be cured.
HINDU—belonging to the religion, Hinduism, which began in India.
MERCHANTS—people who buy things then sell them for a profit.
MINED—dug out of the Earth.
MINERAL—a natural substance that is not an animal or a plant.
ORE—a rock or mineral which contains a metal that is valuable enough to be mined.
PHARAOHS—kings of ancient Egypt.
POLLUTE—to spoil or make dirty.
PRECIOUS—something of great value.
PROFIT—the amount of money made when something is sold.
RARE—unusual or uncommon.
REFLECT—to throw back or show an image of.
VALUABLE—worth a lot of money.

Did you know?

The pyramids were built as tombs for the Pharaohs who ruled Egypt thousands of years ago. The Pharaohs were supposed to put a curse on their tombs so that anyone who stole from them would die. The tombs were full of fabulous treasures and many robbers were prepared to risk the curse in order to gain the riches.

Index

According to the fairytale, Jack risked his life climbing the beanstalk to a castle in the sky to steal the most treasured possession of the giant who lived there—the goose that could lay golden eggs. What would you do to get treasure that would last a lifetime?

Did you know?

The drill your dentist uses on your teeth is covered in lots of very tiny diamond chips.

Fact of the day

When gold is heated, it doesn't boil until it reaches 2,660°C (4,820°F). Water boils at 100°C (212°F).